CREATIVE GRAPHING

by Marji Freeman

Cuisenaire Company of America, Inc.
10 Bank Street. P.O. Box 5026
White Plains, NY 10602-5026

Cuisenaire Company of America, Inc.
10 Bank Street, P.O. Box 5026, White Plains, NY 10602-5026

ISBN O-914040-47-2

Printed in the United States of America

7 8 9 10-BK-98 97

Creative Graphing

Yancey

For decades teachers have searched for ways to motivate their students. As a math teacher for thirteen years and a math consultant for the past five years, I have learned that one of the easiest ways to motivate students is to get them involved with their mathematics. This means taking mathematics out of the textbook and putting it into real-life situations. With this approach students are willing to work problem after problem because it is interesting, relevant to their lives and not the traditional worksheet or textbook page approach to "doing" math.

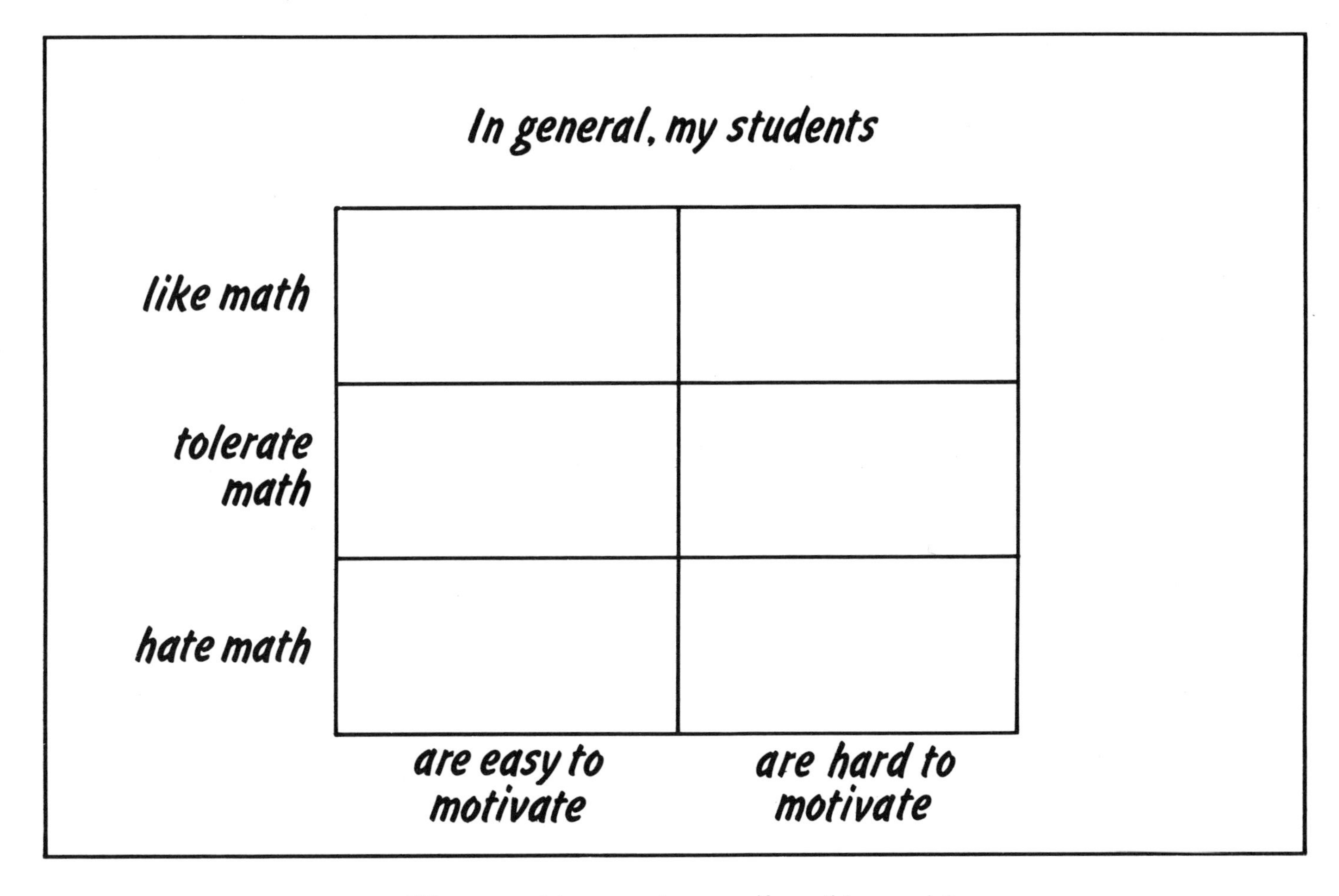

Where would you put yourself on this graph?

If you put your "X" in the top left-hand corner you probably don't need this book, but I think you'll find it interesting anyway. But, if like most of us math teachers, your "X" fell into one of the other rectangles, this book was made for you.

While this book is not a cure-all, it is a different way to give meaning to mathematics. The graphs you find in this book are different from the graphs found in a textbook. The graphs in a textbook already have data on them. The graphs you will be using must be responded to by the students, thereby providing data that is particular to your students. Students will be learning about each other as well as mathematics from the graphs.

This book is divided into four sections. The first section deals with the details of why? when? how? to use graphs. The second section discusses the most important part of the graphing experience–processing. The third section provides examples of graphs you might use with teacher notes for some of the graphs. The fourth section contains additional ideas for graphs.

I hope you and your students will have as much fun with graphing and mathematics as my students and I have had.

Why? When? How?

Why do I use graphs?

Graphs present information visually, making data and relationships among data clearer than information presented other ways. Students need opportunities to deal with graphs. After all, we are inundated with graphs everyday in magazines and newspapers. Graphs provide variety and a means for problem solving. By building, interpreting and summarizing graphs, students are helped to develop their critical thinking skills.

When I use graphs in my class, I find students to be:

- motivated
- ready to solve problems in relation to the graph
- able to read and interpret other graphs and tables
- capable of organizing data
- able to summarize information
- willing to predict
- better at estimating
- willing to do more mental calculations
- able to solve word problems

How often should I use graphs?

Graphs should be used as often as you like. However, the more they are used, the better the students become at interpreting graphs and solving mathematics problems. I use a graph about once a week. If something special is happening in a week I might use an additional graph. For instance, when MASH was showing its final episode, we had a graph that asked "Did you watch MASH last night?" Around Superbowl time, an appropriate graph asks "Who do you want to win the Superbowl?"

How should I manage students recording on graphs?

I like to draw my graphs (or a student does) in advance and laminate them. I either use butcher paper or large graph paper that I have purchased. Students respond on the graph using a grease pencil, overhead projector pen or adhesive colored dots. This way, when we have done the graph, I can clean it off and it is ready to be used the next year. (If you use the adhesive dots, they need to be removed within a day or two or they are difficult to remove.) Sometimes I sketch a graph on the chalkboard or on an overhead transparency. Those graphs get erased when we are finished with them.

I usually have a graph up for my students on Monday morning. Since I teach middle school, I use one graph for all my classes. All day long students fill in the graph. It is interesting to watch the data grow throughout the day. If we want to know how each class responded, we use a different color of dot or pen for each class or each class has a separate graph.

I have tried several ways of managing how students record on the graph. I have had students record as they come into class, but I found that to be too chaotic. Then I tried having students respond to the graph after roll check, a few at a time. This I found to be too noisy because students were waiting with nothing to do. Finally I discovered that the best method for me was to wait until we had gone through our normal routine and students were working on their assignments. Then a few students at a time go up to the graph and respond to it. By the end of the day all classes have responded to the graph and it is then ready to be processed sometime later during the week.

The elementary teacher who sees the same students all day will not have to wait until the end of the day to have all the data for processing. If you teach in a situation where you pull out students from a class, have the students respond to the graph and save the data for processing at another time. The "floating" teacher may want to use the idea of a transparency that is easy to take to the next classroom.

What do I do now?

During the week we take time to summarize the data and solve math problems using the data. (Examples of questions can be found in the "Processing" section.) I usually leave the graph up all week so that we can refer to it now and then. During the week, I find students discussing the data on their own, often making their own predictions or analyzing why the data looks the way it does.

When should I use graphs?

Along with using graphs once a week, I use graphs when a particular need arises. For instance, for a PTA presentation, our Math Department Chairperson needed to give an estimate of how much homework time parents could expect their children to have each night. In order to supply the Department Chairperson with real data, I put the following graph up.

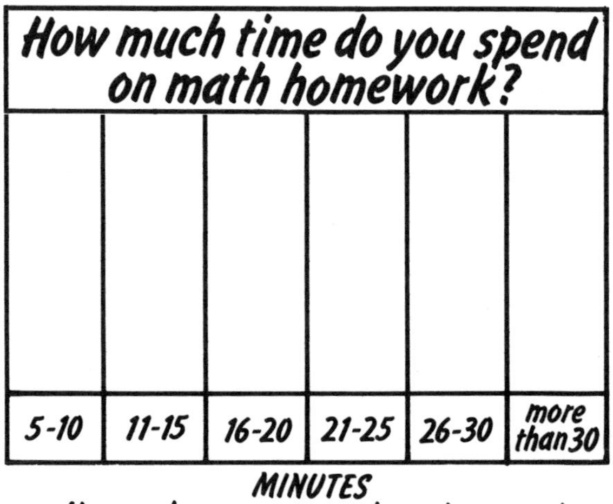

How much time do you spend on math homework?

5-10	11-15	16-20	21-25	26-30	more than 30

MINUTES

Use a dot to respond to the graph.

My students recorded on it. The following day as a class we determined an estimate of the average amount of time students might spend on their homework. I turned in our estimate feeling confident that I had turned in a realistic amount of time and not just an arbitrary number of minutes.

I have used graphs in helping me make lesson plans. Through a graph I can ask my students their opinions of different class procedures or topics.

After having used graphs in my classes, I decided I needed to know how my students felt about graphs. So I put this graph up in order to find out.

Do you like to respond to graphs?

Do you feel you learn from responding to graphs?	Yes	No
No		
Yes		

I did have a few students say they didn't like responding to graphs, but they still felt they had learned from them. About 90% of my students liked graphs and felt they learned from them. We discussed what kinds of things they had learned from graphs. We discussed ways to change graphs to improve them. Based on this data and the discussion that followed, I chose to continue the graphs in my classroom.

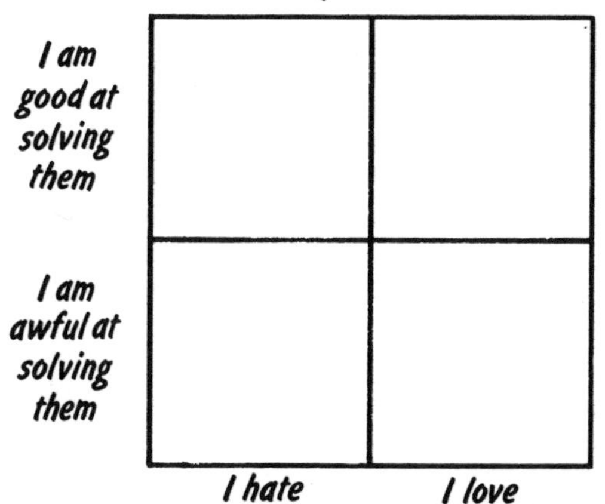

The answers on this graph tended toward the "I hate" and "I am awful at solving word problems." We discussed the fact that quite a few students did not like word problems and felt unsuccessful with them. We analyzed why students might feel that way. Students offered that they did not understand word problems and that they needed to do more and talk about them in order to get better at solving word problems. The general conclusion was that we should do more word problems in class. The graphs are ideal for providing word problem situations for students to discuss and solve. The "Processing" section describes many such examples.

Where do I get ideas for graphs?

One way I get ideas for graphs is to think of a category. For instance, I might think of "names" as a category. Then I start thinking about questions dealing with names:

- If you could, would you change your first name?
- Were you named for someone?
- Is your first name longer, shorter or the same length as your last name?
- Do you have a middle name?
- Do you have a nickname?

Now I have five questions for topics for graphs. You will find ideas for graphs by category in the last section of the book.

Another source for topics for graphs is the national, state, local and campus news.

- Do you think there should be a seatbelt law?
- Who do you want to be President?
- Are you going to the Valentine Dance?

Another good source for graphing ideas is the students. In groups of four I have students brainstorm ideas for graphs. I ask them to think of things they would like to know about their classmates. They sketch four or five of their best ideas. Each group then gives me the title of their favorite or best graph. I keep track of topics chosen so that we do not repeat a topic. Each group then draws their graph on graph paper. I have the graphs laminated and then we use them in class. I spread this activity over several days:

Day 1 spend 10-20 minutes brainstorming ideas for graphs

Day 2 spend 10-20 minutes adding to the list and choosing 4 or 5 best topics, turn these in to the teacher

Day 3 spend 10-20 minutes coming to a concensus as to which graph the group wants to do

Day 4 spend the period drawing the graph on graph paper

Processing

The most important part of using graphs in the classroom is to process them thoroughly. It is from this exploration that students have the opportunity to examine the data, analyze the information, and interpret the results.

There are several ways I manage processing. One way is to ask questions orally. Students raise their hand and answer the question. Sometimes students discuss the questions in small groups before giving an answer. Other times I ask the questions and students reply in writing, either on their own or in small groups.

The types of questions I ask fall into three categories.

Basic Skills Questions

These are questions that require that the students use basic number skills in answering, adding, subtracting, multiplying, dividing whole numbers, decimals, or fractions, or using percents, or comparing numerical data. If we are looking for an exact answer, then there is only one right answer with this type of question. If the question asks for an estimate, then a range of answers is acceptable. The main purpose of this type of question is to allow students to apply their basic number skills in the context of real information. In this way, students are really solving word problems.

Analytical/Extension Questions

These are questions that encourage students to analyze and summarize the information on the graph. Students "extend" the graph by predicting other possible outcomes and analyzing how they arrived at their answers. There usually can be more than one right answer with this type of question.

Personal/Opinion Questions

These questions allow the student to express an opinion or make a judgment about a situation. With this type of question, all answers are acceptable.

I have used these three categories of questions in a variety of ways when dealing with graphs:

- Sometimes a graph seems to warrant only 5-10 minutes of processing, while at other times we might spend 30 minutes answering questions orally or in small groups.
- Sometimes I write several questions on the board or on a worksheet for students to answer working individually or in small groups.
- Sometimes I have the students make up the questions. They either exchange questions or I include them all on a worksheet for all students to try.

Some graphs lend themselves to extensive processing while others do not. However, no matter how much time is spent on questioning, I try to include some from each of the three categories of questions. In this way not only have we had practice with a basic number skill and/or estimation, but we have extended our thinking.

Sample Graph and Questions

Below is an example of a graph that already has data filled in. The graph is followed by examples of questions in each of the three categories. Examine the questions and decide which are applicable to your grade level.

What is your favorite night for watching TV? Mark an 'X'.

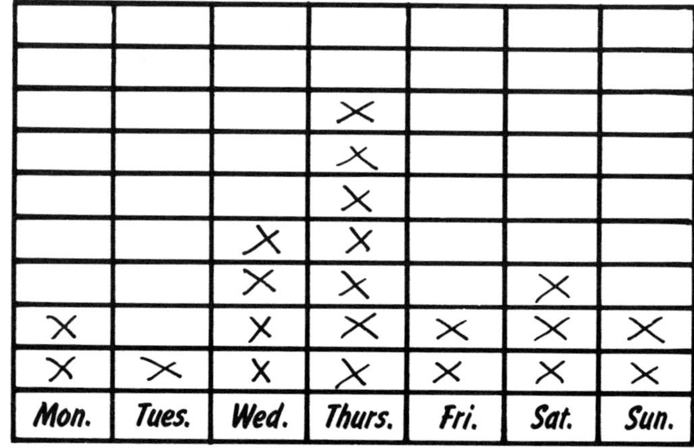

Days of the week

Basic Skills Questions

Remember: You will be dealing with real data, so fractions and percents will not necessarily come out even. It is important for students to realize that answers in the real world are not exact all the time and that estimates can be appropriate.

1. Which night had the most votes? Which night had the least number of votes?
2. Which nights tied in their number of votes?
3. How many more people prefer their night to Sunday night?
4. List the nights in order from most votes to least votes?
5. Which night has two less votes than Wednesday's votes?
6. How many people prefer Friday, Saturday or Sunday nights? (This is a question dealing with union of sets. Do students realize that "or" means they have all three together?)

7. How many people are on the graph? (How did you find out how many people are on the graph? Did anyone find out another way?)
8. What fraction of the votes are on Monday night? (Tuesday, Wednesday, etc.)
9. About what percent of the people prefer Saturday night?
10. About what percent of the people prefer a week night? (How did you decide?)
11. Is there a night that has 1/7 of the votes? How did you decide?

Analytical/Extension Questions

1. What do we know about this group as far as their favorite nights for watching TV are concerned?
2. What TV programs are found on the favorite night(s)?
3. Do you think this information would be the same if we polled all (grade level you teach) 7th graders in our school? Why or why not? How about adults? Why or why not?
4. What are some reasons that more people chose Thursday as their favorite nights for watching TV?
5. Who would be interested in or need this kind of information?
6. What are some other questions we could ask about this graph?

Personal/Opinion Questions

1. What influences TV watching?
2. What did you like/dislike about this graph? Rate this graph with other graphs we have done according to appeal and information.
3. Why did you choose your favorite night for watching TV?
4. Why do you think someone chose Sunday (or another night you did not choose) as their favorite night?
5. Do the results of this graph seem reasonable to you?
6. Should students be limited in the amount of TV they can watch? Explain.
7. Should students be limited in the types of programs they can watch? Explain.
8. What type of programs do you prefer? Why?

A few last notes before you begin graphing...

- Be sure to model for the students how to record on the graph if they have never responded to the type of graph you have posted.
- When all data is on the graph, ask questions from each of the three categories. Some suggested questions accompany the graphs that follow. Some of the questions are best asked in sequence. For instance, "What percent of the people prefer Thursday for watching TV?" You may want to begin with "How many people prefer Thursday?", then "What percent of the people prefer Thursday?".
- The questions I have suggested need to be rewritten to fit the data that is actually collected. For instance, with the newspaper graph I ask, "How many more people read the paper four times than three times?". If in your data nobody reads the paper four times a week, you will need to change that number.
- Additional activities are suggested for each graph. These ideas will help in integrating other curriculum areas.
- Many of the graphs I have suggested have just the vertical and horizontal axis drawn without lines drawn. You may want to draw the lines or use graph paper which already has the lines drawn.

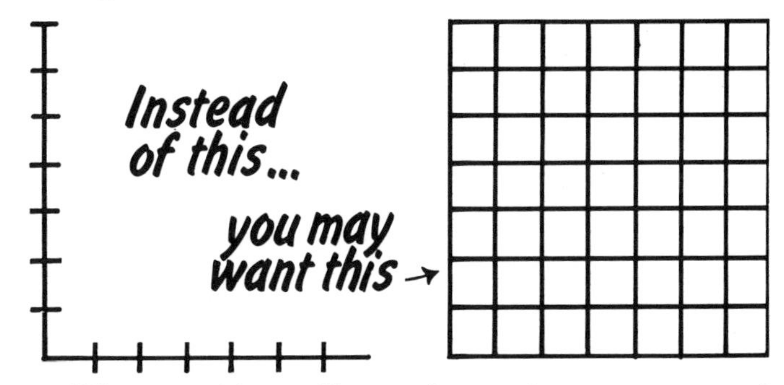

- When working with graphs, make sure you and your students summarize the data. Answer the question "What do we know about our class as a result of this graph?".
- Sometimes, before having students respond on a graph, ask them to predict what they think the results will be.
- Several of the graphs use Venn Diagrams. A Venn Diagram graph lends itself to the discussion of union of sets, intersection of sets, empty sets, universal sets. I have found that most students have difficulty understanding a Venn Diagram at first. If this is new for your students, discuss what the interesting circles mean before the students are to respond on a graph.

Teacher Notes

Do you listen to a radio or TV when you do your homework?

BASIC SKILLS QUESTIONS

- About what percent of the students watch TV while doing their homework?
- What fraction of the students don't watch TV or listen to the radio?
- How many students listen to the radio while doing homework?

ANALYTICAL/EXTENSION QUESTIONS

- Why do some students listen to the radio instead of watch TV while doing homework?
- What do we know about this group's study habits?
- Compare and contrast this graph with others we have done.

PERSONAL/OPINION QUESTIONS

- Should students watch TV/listen to the radio while doing homework? Explain your answer.
- Is the information we have important? Explain. Who would need or want to have this information?

OTHER ACTIVITIES

- Poll the students in your classroom or grade level to find out what the favorite TV station is.
- Research the effects that watching TV/listening to radio while doing homework might have on grades/learning.

NOTE: Students who don't watch TV or listen to the radio should indicate by putting their dot outside the circles.

Are you the oldest, youngest or in-between child?

BASIC SKILLS QUESTIONS

- How many students responded on this graph?
- Give the decimal that represents the number of oldest students.
- What percent of the students are the youngest child?

ANALYTICAL/EXTENSION QUESTIONS

- Brainstorm reasons it might be easier to be the oldest child? the youngest child?
- List some advantages/disadvantages to being the youngest child, in-between child, oldest child.
- What are some other questions we could ask about this graph?

PERSONAL/OPINION QUESTIONS

- If you had a choice, would you rather be the youngest, the oldest or the in-between child? Explain your answer.
- Do you think parents tend to treat the oldest child differently from the way they treat the youngest child? Why?

OTHER ACTIVITIES

- Find out if your classmates' parents are an oldest, youngest or in-between child. Graph the information.

NOTE: I have my students who are an "only" child count themselves as an oldest child.

Do you listen to a radio or TV when you do your homework?

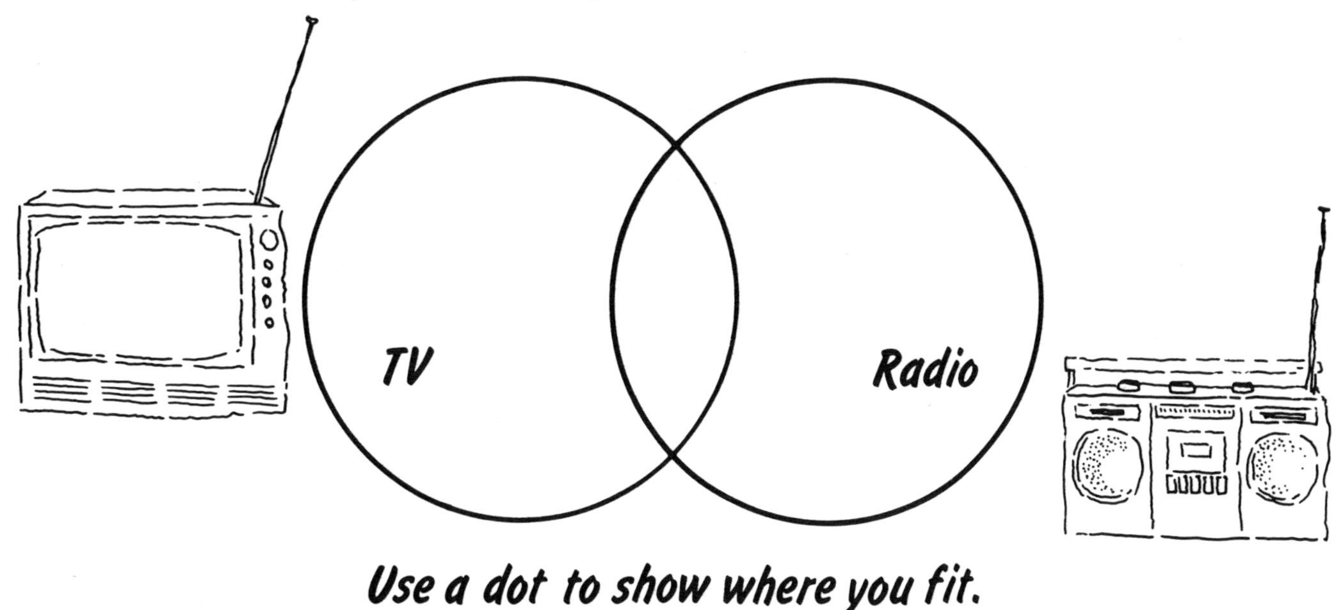

Use a dot to show where you fit.

Are you the oldest, youngest or in-between child?

Oldest	Youngest	In-between

Use a dot to show where you fit.

Teacher Notes

What is your favorite way to spend a Saturday?

BASIC SKILLS QUESTIONS

- What fraction liked to spend Saturday with friends?
- How many people prefer to spend Saturday alone than with friends? family?
- Which way of spending Saturday was the least popular?

ANALYTICAL/EXTENSION QUESTIONS

- What are reasons why a person would like to spend Saturday alone?
- How do you think parents would answer this graph?
- What are some activities you might do with family on Saturday?

PERSONAL/OPINION QUESTIONS

- What kinds of things do you do on Saturday?
- Ten years from now, what kinds of things do you think you will be doing on Saturday?
- Describe a "perfect" Saturday.

OTHER ACTIVITIES

- Predict how responses would differ if the graph asked about Sunday. Post a graph about Sunday to find out.

Do you wear a seat belt when riding in a car?

BASIC SKILLS QUESTIONS

- What percent of the students who answered the graph always wear a seat belt?
- How many people thought a person shouldn't wear a seat belt?
- What fraction of the people thought you should wear a seat belt but never wear one?

ANALYTICAL/EXTENSION QUESTIONS

- What reasons do people have for not wearing a seat belt?
- Do you think the answers to this graph would be different if done by your parents?
- Who would need this information?

PERSONAL/OPINION QUESTIONS

- Should it be against the law for people not to wear a seat belt? Explain your answer. What should be the punishment for a person who doesn't wear their seat belt?
- Do you think you will wear a seat belt when you drive? Why or why not?

OTHER ACTIVITIES

- Research how many states have a seat belt law?
- Gather statistics on accidents to see if people are better off wearing seat belts.

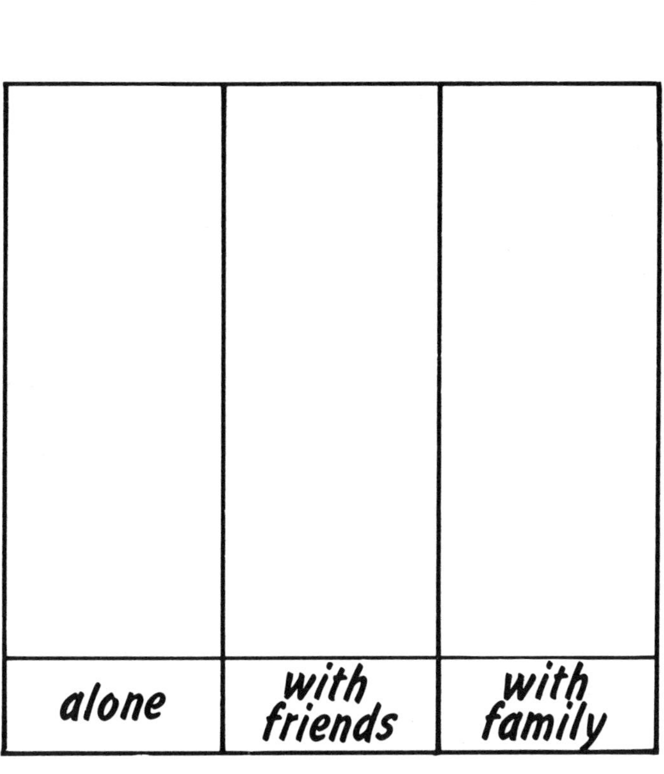

Use a dot to show where you fit.

What is your favorite way to spend a Saturday?

Do you wear a seat belt when riding in a car?

Should a person wear a seat belt?

	always	usually	sometimes	never
yes				
no				

Mark an 'X' where you belong.

Teacher Notes

What is your favorite cartoon character?

BASIC SKILLS QUESTIONS
- Which was the favorite cartoon character?
- What percent of the students liked Garfield as their favorite cartoon character?
- What fraction chose another cartoon character other than those listed?

ANALYTICAL/EXTENSION QUESTIONS
- What are some other cartoon characters that could be put on this graph?
- Who would need to know the information we have gathered?
- What do we know about our class and their favorite cartoon character?

PERSONAL/OPINION QUESTIONS
- What is your favorite cartoon character? Why?
- Do you like cartoon characters that are "human" or "animal"? Why?

OTHER ACTIVITIES
- Create your own cartoon character.
- Find out about your favorite cartoon character. Who created him? How old is he? etc.

Where were you born?

BASIC SKILLS QUESTIONS
- What percent of the students were born in Texas but not in San Antonio?
- How many were born in San Antonio?
- Write the decimal that indicates how many were born in the U.S.A.

ANALYTICAL/EXTENSION QUESTIONS
- What do we know about this group from the graph? (Summarize.)
- Create another graph that shows this information.
- What kind of prediction can you make about the entire school as to where the students were born?

PERSONAL/OPINION QUESTIONS
- If you could choose to be born in any state, which one would you choose and why?
- Would you rather live in a small town or a big city? Explain your answer.

OTHER ACTIVITIES
- Poll the class to determine in what city each classmate was born. Make a graph showing this information.

Garfield	
Snoopy	
Smurfs	
Ziggy	
Dennis the Menace	
Other	

What is your favorite cartoon character?

Use a tally mark.

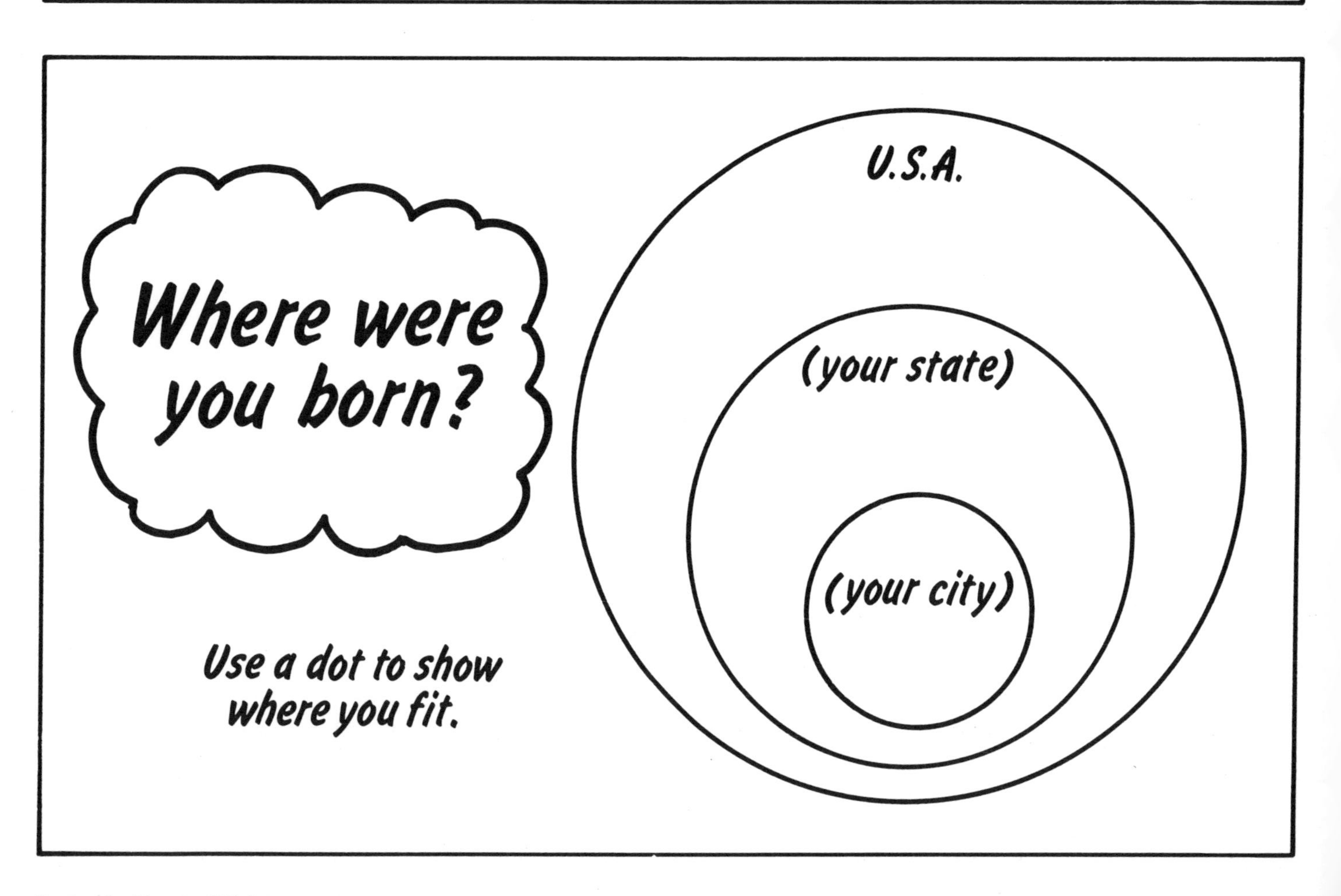

Teacher Notes

Which of these days do you like best?

BASIC SKILLS QUESTIONS

- How many chose a holiday as the day they like best?
- What percent of the students chose the last day of school as their favorite day?
- What fraction said their birthday was their favorite day?

ANALYTICAL/EXTENSION QUESTIONS

- What are some reasons why a student would pick their birthday, Christmas, Easter, etc. as their favorite day of the year?
- How did the day you chose compare to the class?
- What other days could have been put on the graph?

PERSONAL/OPINION QUESTIONS

- Why did you choose the day you chose?
- If you could declare a "holiday", what would it be?
- Is too much emphasis placed on holidays or special days? Explain.

OTHER ACTIVITIES

- Research and report on birthdays or holidays in other lands.

Do you watch cartoons on Saturday mornings?

BASIC SKILLS QUESTIONS

- How many students watch cartoons on Saturday morning?
- What fraction of the students do not watch cartoons?
- Do more students watch cartoons or not watch cartoons?

ANALYTICAL/EXTENSION QUESTIONS

- What are some other things to do on Saturday mornings instead of watching cartoons?
- Would the data on this graph be different if asked of 5-year-olds?
- What are reasons students your age like to watch cartoons? don't like to watch cartoons?

PERSONAL/OPINION QUESTIONS

- Are cartoons good for children to watch? Explain your answer.
- In what ways could Saturday morning TV be improved?
- Should kids be limited to how many hours of cartoons they watch? Why or why not?

OTHER ACTIVITIES

- Create a graph to learn how many hours kids spend on Saturday morning watching cartoons.

Which of these days do you like best?
your birthday
Christmas
Easter
4th of July
New Year's Day
Valentine's Day
First day of school
Last day of school
Use a dot to show where you fit.

Do you watch cartoons on Saturday mornings?
Yes
No
Mark an 'X' where you fit.

Teacher Notes

Where do you fit on this graph?

BASIC SKILLS QUESITONS

- What percent of the people who answered this graph are girls? boys?
- What percent of the boys were born in San Antonio?
- What fraction of the people were not born in San Antonio?

ANALYTICAL/EXTENSION QUESTIONS

- What are some reasons why girls and boys are not living in the city they were born in?
- Do you think most girls and boys are still living in the city they were born in?
- Do you think the results of this graph would be different if the faculty had filled it in? Why or why not?

PERSONAL/OPINION QUESTIONS

- Were you surprised by the results of this graph? Explain.
- Have you lived in other places? Which was your favorite?
- How could you change this graph to get additional or different information?

OTHER ACTIVITIES

- Look up facts about your city.
- Plan a bulletin board or poster about your city.

Which of these fast foods do you like best?

BASIC SKILLS QUESTIONS

- How many more liked hamburgers than hotdogs?
- What percent of the students like tacos as their favorite food?
- What fraction liked Chinese food?

ANALYTICALEXTENSION QUESTIONS

- How many "fast food" places are there in your neighborhood?
- What is meant by the term "fast food".
- Give pros and cons of eating "fast foods".

PERSONAL/OPINION QUESTIONS

- Should only "fast food" be served for lunch in the school cafeteria? Why or why not?
- Do you think "fast food" is healthy? Explain.
- What do you think the main reason is that people eat "fast food".

OTHER ACTIVITIES

- Create a graph to find out how many times a week each student eats "fast food".
- Poll people to find out why they eat "fast food".
- Find out the calorie count and nutritional value of several "fast foods". Compare them to non-"fast foods".

Where do you fit on this graph?

Girl	Boy
Born in	(your city)

Mark an 'X' where you belong.

Which of these fast foods do you like best?

chicken	tacos	fish	hamburgers	chinese	hot dogs	pizza

Use a dot to show where you belong.

Teacher Notes

How do you see yourself?

BASIC SKILLS QUESTIONS

- What percent of the students found themselves to be organized, hard working and creative?
- What fraction felt they were only creative?
- How many felt they were organized and hard working?

ANALYTICAL/EXTENSION QUESTIONS

- What are some other qualities that could be put in the circles?
- What is meant by the terms organized, hard working and creative.
- What would happen if the circles did not overlap?

PERSONAL/OPINION QUESTIONS

- Which one of these qualities do you feet is the most important? Why?
- What other positive qualities do you feel you have?
- If you could add another quality to the graph, what would it be? Explain.

OTHER ACTIVITIES

- Find other words that mean organized, hard working, creative.

How do you get to School?

BASIC SKILLS QUESTIONS

- What percent of the students only come to school by motorized vehicles?
- How many students come to school in two different ways?
- What is the most common way of coming to school?

ANALYTICAL/EXTENSION QUESTIONS

- What are some of the factors that determine how students come to school?
- Do you think the data on this graph would be different for elementary students? high school students? college students?
- Who would need the kind of information on this graph?

PERSONAL/OPINION QUESTIONS

- How will the ways you come to school change as you get older?
- Should schools be close enough so that everyone can walk or ride their bike to school?
- What is your favorite way to come to school?

OTHER ACTIVITIES

- Create a method for determining how the entire school comes to school (questionnaire, actually observe how students come to school, etc.).

How do you see yourself?

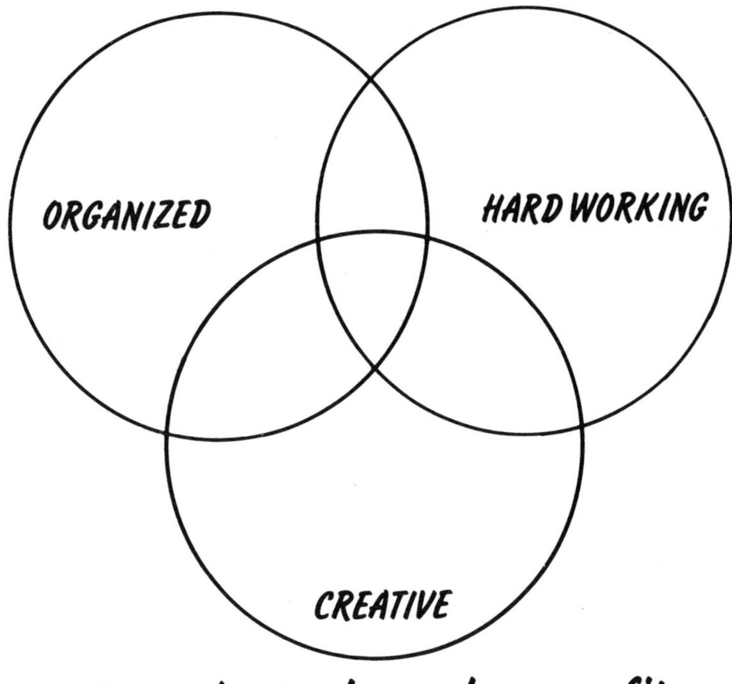

Use a dot to show where you fit.

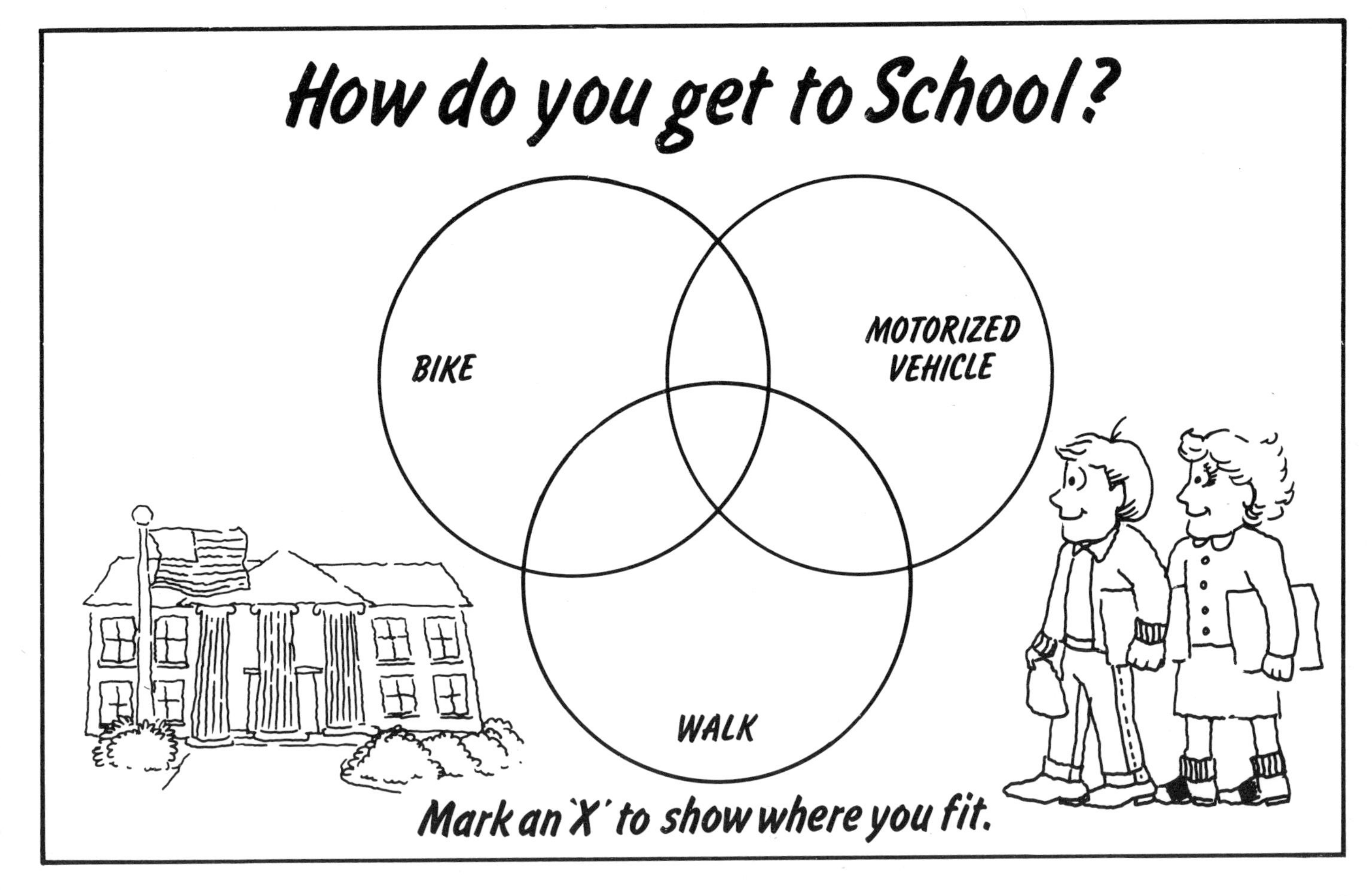

Teacher Notes

OUR CLASS

BASIC SKILLS QUESTIONS

- How many in the class have at least one brother?
- What percent of the class has lived in Texas for 5 years or more?
- Did more members of the class have a hobby or take a vacation this summer?

ANALYTICAL/EXTENSION QUESTIONS

- How is this graph different from other graphs you have done?
- Determine a set of characteristics to be used in this graph for adults. Could you keep some of these characteristics?
- Summarize the graph.

PERSONAL/OPINION QUESTIONS

- Do you think it's important or interesting to know about your classmates? Explain.
- What are other characteristics you would add to this graph?

OTHER ACTIVITIES

- Use the same graph with another class. Compare the results of your class graph to the other class.
- Create a new graph with different characteristics.

NOTE: Each student responds to each of the characteristics by placing an "X" or a dot by each of the characteristics that he/she possesses.

Does your family consist of brothers, sisters, or both?

BASIC SKILLS QUESTIONS

- How many students have at least one sister?
- What percent of the class only has at least one sister?
- How many students answered this graph?

ANALYTICAL/EXTENSION QUESTIONS

- What would happen if the circles did not overlap (disjoint sets)?
- How else could we have gotten this information?

PERSONAL/OPINION QUESTIONS

- How many children are in your family? How many do you think is ideal? Explain.
- What qualities do you like in your brother? sister?
- How do you try to be a good brother or sister?

OTHER ACTIVITIES

- Poll the class to see exactly how many brothers and sisters each student has and how many have no brothers or sisters. Put this information in a graph.

OUR CLASS

has brown hair											
has at least one brother											
born in January											
has a hobby											
took a vacation this summer											
has lived in (your state) 5 years or more											
likes math											

Write your initials on one of the blanks. Mark an 'X' for each of the characteristics you have.

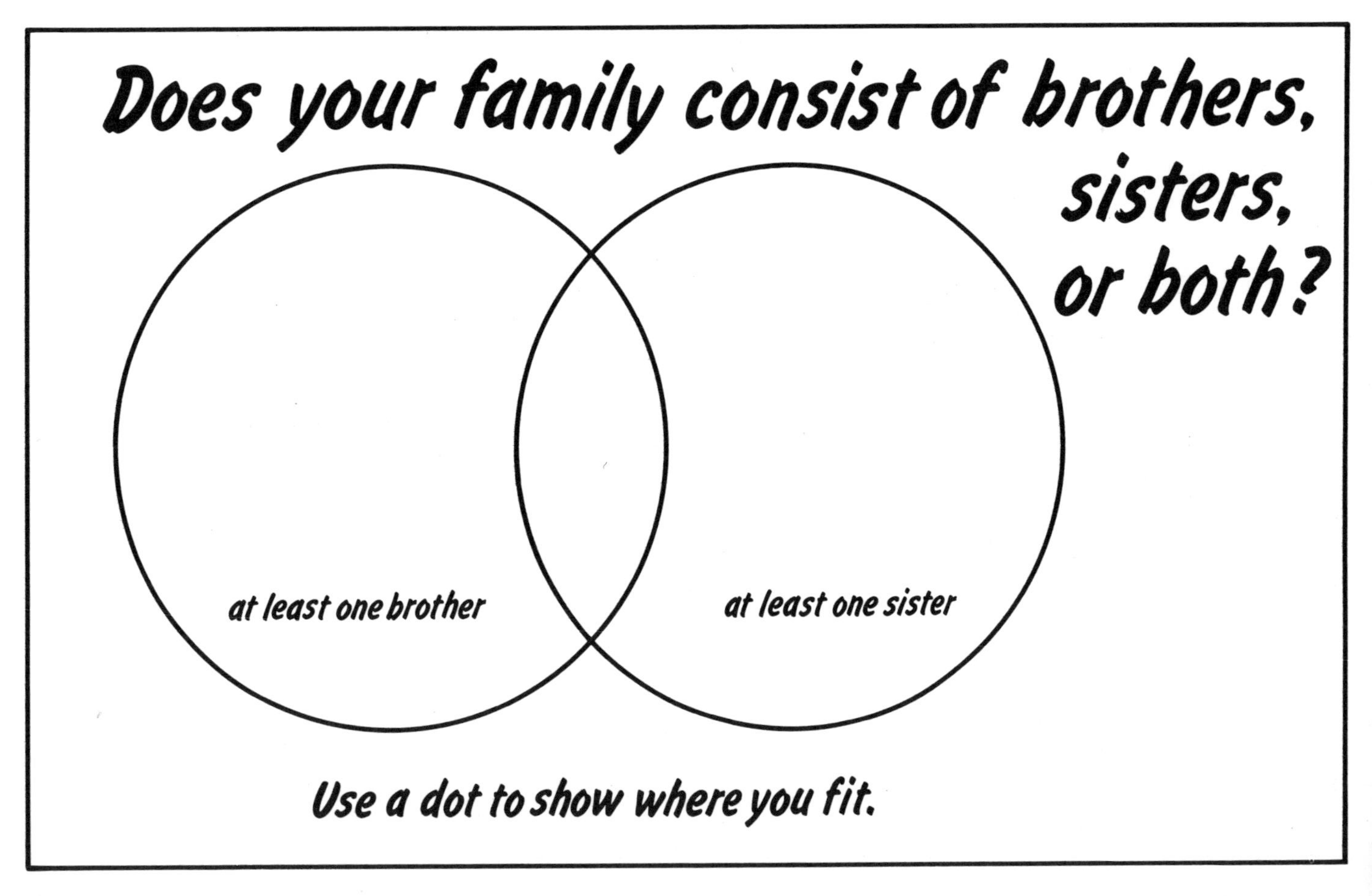

Teacher Notes

Do you have a pet?

BASIC SKILLS QUESTIONS
- How many students have a pet?
- What percent of the students have a pet?
- Write the fraction that represents the students that don't have a pet.

ANALYTICAL/EXTENSION QUESTIONS
- What kinds of pets can a person have in the city?
- What are some reasons why a person might not have a pet?
- What are some ways pets are helpful to people?

PERSONAL/OPINION QUESTIONS
- If you had a choice, what kind of pet would you have?
- How would you decide what to name your pet?
- Should people have pets? Explain.

OTHER ACTIVITIES
- Prepare to instruct younger students on how to care for their pets.

When I think about what my career might be...

BASIC SKILLS QUESTIONS
- What percent of the students have several ideas about their career?
- How many have no ideas about what their career will be?
- What fraction think they know what their career will be?

ANALYTICAL/EXTENSION QUESTIONS
- What factors should you consider in determining what career to pursue?
- In what careers do you use math?
- Construct another graph that asks for the same information.

PERSONAL/OPINION QUESTIONS
- Why is it hard to decide on a career?
- At what age should a person decide what their career will be?
- What career are you thinking of going into? Why? What kind of education/training will you need?

OTHER ACTIVITIES
- Choose a career you are interested in to research and learn about. Present your findings to the class.

When I think about what my career might be...

I find I have no ideas.	I find I have several ideas.	I think I know what it will be.

Use a dot to show where you fit.

Teacher Notes

What will your Halloween mask look like?

BASIC SKILLS QUESTIONS
- What percent will wear a funny mask?
- How many will wear some other kind of mask?
- What fraction will wear a pretty or funny mask?

ANALYTICAL/EXTENSION QUESTIONS
- Organize this information in some other way.
- Contrast this graph with others we have done.
- How can we determine how many people answered the graph?

PERSONAL/OPINION QUESTIONS
- How do you decide what mask to wear on Halloween?
- Do you go trick-or-treating? Why? Why not?

OTHER ACTIVITIES
- Make a list of safety rules for Halloween. (As a class project, adopt a "younger" class and acquaint them with Safety at Halloween.)
- Draw a Halloween mask you would like to have.

In which month were you born?

BASIC SKILLS QUESTIONS
- In what month were the most people born?
- What percent of the people were born in the summer (June, July, August)?
- What fraction of the people were born in March?

ANALYTICAL/EXTENSION QUESTIONS
- Who would use this kind of information?
- What are some birthday traditions that people observe?
- What are other questions we could ask about this graph?

PERSONAL/OPINION QUESTIONS
- If you could choose a different month for your birthday, what month would you choose? Why?
- What do you think would be the worst month to have a birthday? Why?
- What traditions does your family observe during birthdays?

OTHER ACTIVITIES
- Research how birthdays are celebrated in other countries. Compare their traditions to your traditions.

What will your Halloween mask look like?

scary	funny	pretty	other

Mark an 'X' where you belong.

In which month were you born?

Jan.	Feb.	Mar.	Apr.	May	June	July	Aug.	Sept.	Oct.	Nov.	Dec.

Put an 'X' where you belong.

Teacher Notes

If I could change my age, I would be...

BASIC SKILLS QUESTIONS
- How many people want to be the same age?
- What percent of the people want to be older?
- How many more people want to be younger than older?

ANALYTICAL/EXTENSION QUESTIONS
- Would the results be the same if we asked adults?
- How is this graph different from other graphs?
- How can we find out how many people are on the graph?

PERSONAL/OPINION QUESTIONS
- Why would people want to be older? younger? the same?
- What age would you want to be? Why?

OTHER ACTIVITIES
- Poll the faculty (or some grade level) to find out if they would want to change their ages.
- Find out what the oldest age a person has lived to. Would you want to live to be that old? Why? Why not? Under what conditions?

What is the most important quality in a friend?

BASIC SKILLS QUESTIONS
- Which quality received the most votes?
- How many people voted for "honesty" and "cheerful"?
- What fraction of the people voted for intelligence?

ANALYTICAL/EXTENSION QUESTIONS
- What are other qualities that could have been used?
- What do we know about our class concerning important qualities in friends?
- How can we find out which quality received the most votes?

PERSONAL/OPINION QUESTIONS
- Which is the most important quality to you? Explain.
- Why do you suppose people chose "dependable"? (Substitute any quality.)
- How would you change this graph?

OTHER ACTIVITIES
- Take the 3 top vote-getters of these qualities and poll the class to see which one comes out on top. Did the same quality come out number one as before?

If I could change my age, I would be...

older	
the same	
younger	

Use a tally mark to show where you belong.

What is the most important quality in a friend?

Honest	
Loyal	
Cheerful	
Good Listener	
Intelligent	
Dependable	

Make a tally mark.

Teacher Notes

Which of these do you like best?

BASIC SKILLS QUESTIONS

- How many chose strawberry as the ice cream they like best?
- What percent of the students like vanilla ice cream?
- How many more liked chocolate than liked vanilla?

ANALYTICAL/EXTENSION QUESTIONS

- What are some different ways to eat ice cream?
- How many different flavors of ice cream do you think there are? As a group name as many as you can.

PERSONAL/OPINION QUESTIONS

- What did you like/dislike about this graph? How would you change the graph?
- What is your favorite dessert other than ice cream?

OTHER ACTIVITIES

- Research ice cream as to its calorie count and nutritional value.

Where and with whom will you celebrate Thanksgiving?

BASIC SKILLS QUESTIONS

- How many will spend Thanksgiving out of town with family/friends?
- Will a greater percent spend Thanksgiving in town with immediate family or out of town with immediate family?
- What fraction of those who answered the graph will spend Thanksgiving in town with family/friends?

ANALYTICAL/EXTENSION QUESTIONS

- How did you determine how many people will celebrate out of town?
- What are other questions we could ask? What would each quadrant say?
- Compare this graph to others we have done.

PERSONAL/OPINION QUESTIONS

- What do you like/dislike about this graph?
- How does your family celebrate Thanksgiving?
- Why do you think we celebrate Thanksgiving?

OTHER ACTIVITIES

- Research the history of Thanksgiving. Why do we celebrate it?

Which of these do you like best?

Mark an 'X' where you belong.		
Vanilla	Strawberry	Chocolate

Where and with whom will you celebrate Thanksgiving?

	immediate family	family & friends
in town		
out of town		

Put a dot where you belong.

Teacher Notes

How many times a week do you read a newspaper?

BASIC SKILLS QUESTIONS

- How many people read the paper 2 times a week? 0 times?
- How many more read the paper 4 times than 3 times?
- What percent of the class reads the paper at least 2 times a week? Exactly 5 times a week? More than 3 times a week?

ANALYTICAL/EXTENSION QUESTIONS

- How can we find out how many times a week the parents of the students in the class read the paper? the teachers? the 6th grade class?
- Who would need to know this information?
- How is this graph different from other graphs we've answered?

PERSONAL/OPINION QUESTIONS

- Is it important that people read the newspaper? Explain.
- Do you like to read the newspaper? Why? Why not?
- If you were to write an article for a newspaper, what would it be about?

OTHER ACTIVITIES

- Invite a newspaper person to come in and discuss the mathematics of newspaper work. Poll to find out the favorite newspaper of the class.

What is your favorite part of the newspaper to read?

BASIC SKILLS QUESTIONS

- What fraction of the class prefers the comics?
- How many people do not read the newspaper?
- How many people read the newspaper from 6 to 7 times?

ANALYTICAL/EXTENSION QUESTIONS

- How many people read the paper 4 times a week? How do you know that? (the students should realize they can't answer this question because they answered in intervals instead of exact times)
- How is the graph different from the other graphs on reading newspapers?
- What other parts are there to the newspaper?

PERSONAL/OPINION QUESTIONS

- What is your favorite part of the newspaper?
- How often do you think students should read a newspaper?

OTHER ACTIVITIES

- Find the average length of a paragraph, article, etc.

NOTE: The teacher should discuss how to answer this graph if the students have never answered a 2 variable graph in class. I usually tell my students that they have two questions to answer but can put only one mark on the graph. I have them decide what their favorite part of the newspaper is and then how many times a week they read a newspaper; and like a times table chart, where ever those two answers meet is where their mark goes.

How many times a week do you read a newspaper?

7 ______________________

6 ______________________

5 ______________________

4 ______________________

3 ______________________

2 ______________________

1 ______________________

0 ______________________

Make a tally mark.

What is your favorite part of the newspaper to read?

How many times a week do you read a newspaper?

	National, State News	Local News	Sports	Entertainment	Comics	I do not read the newspaper
6 – 7						
3 – 5						
0 – 2						

Use a dot to show where you fit.

Teacher Notes

Write questions below that would be appropriate for your class for the graphs on the next page.

Which do you prefer?

BASIC SKILLS QUESTIONS

-
-
-

ANALYTICAL/EXTENSION QUESTIONS

-
-
-

PERSONAL/OPINION QUESTIONS

-
-
-

OTHER ACTIVITIES

-
-
-

If you could watch TV only one night a week, which night would you choose?

BASIC SKILLS QUESTIONS

-
-
-

ANALYTICAL/EXTENSION QUESTIONS

-
-
-

PERSONAL/OPINION QUESTIONS

-
-
-

OTHER ACTIVITIES

-
-
-

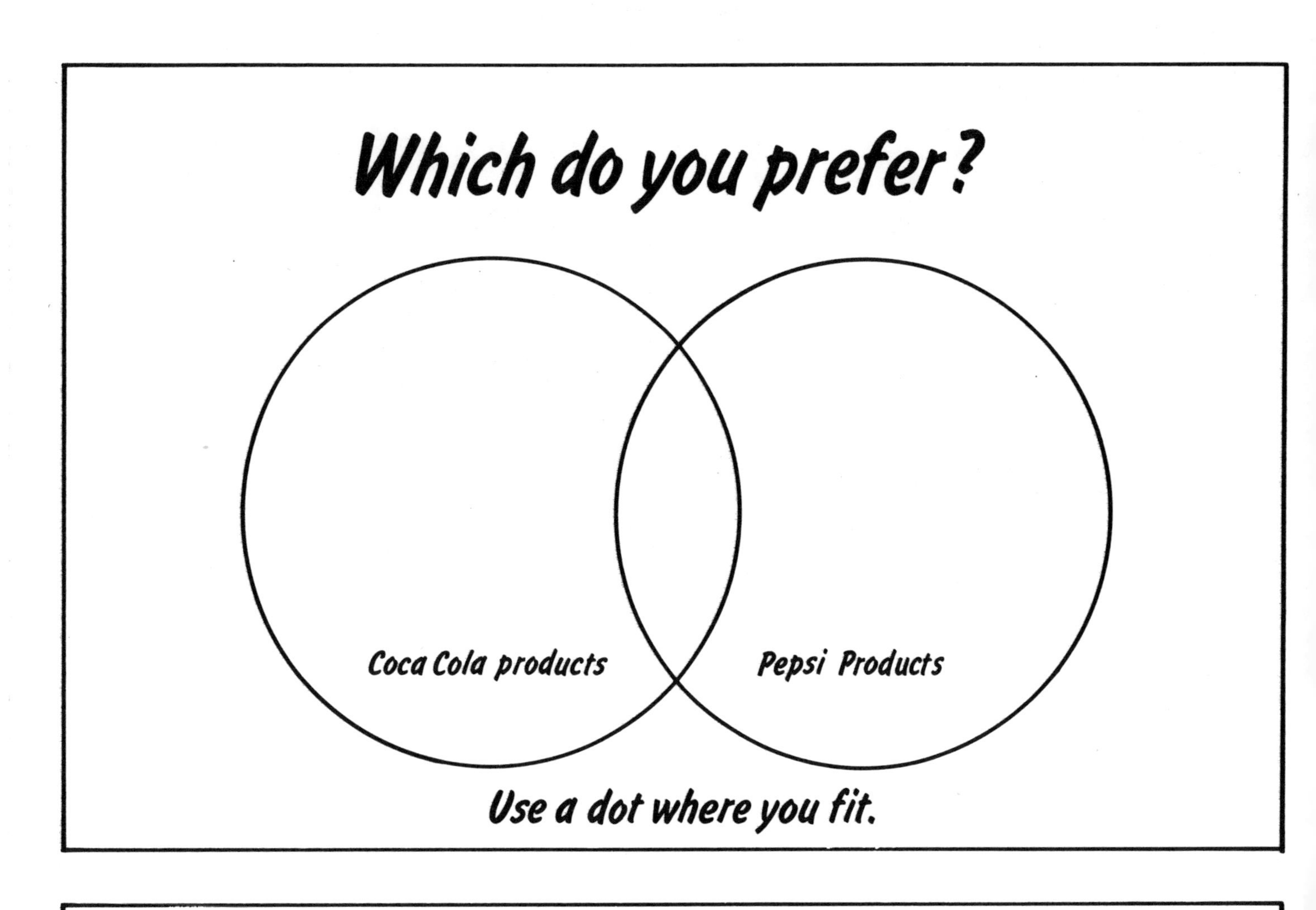

If you could watch TV only one night a week, which night would you choose?

Sunday	
Monday	
Tuesday	
Wednesday	
Thursday	
Friday	
Saturday	

Use a tally mark.

Graphs about Names

Do you prefer to be called by a nickname?

Use a tally mark.

yes ____________________

no ____________________

Is your first name longer, shorter or the same length as your last name?

longer ____________________

shorter ____________________

the same ____________________

Use a tally mark.

Do you have a middle name?

yes	no

Mark an 'X'.

Were you named for someone?

yes	no

Use a dot to build a bar graph.

If you could, would you change your first name?

yes	no

Use a dot.

How many letters are in your first name?

Use a tally mark.

1-5 ____________

6-10 ____________

11-15 ____________

more than 15 ____________

How many vowels are in your first name?

1 ____________

2 ____________

3 ____________

4 ____________

more than 4 ____________

Use a tally mark.

Where does your initial fit in the alphabet?

A-I ____________

J-R ____________

S-Z ____________

Use a tally mark.

Graphs about Breakfast

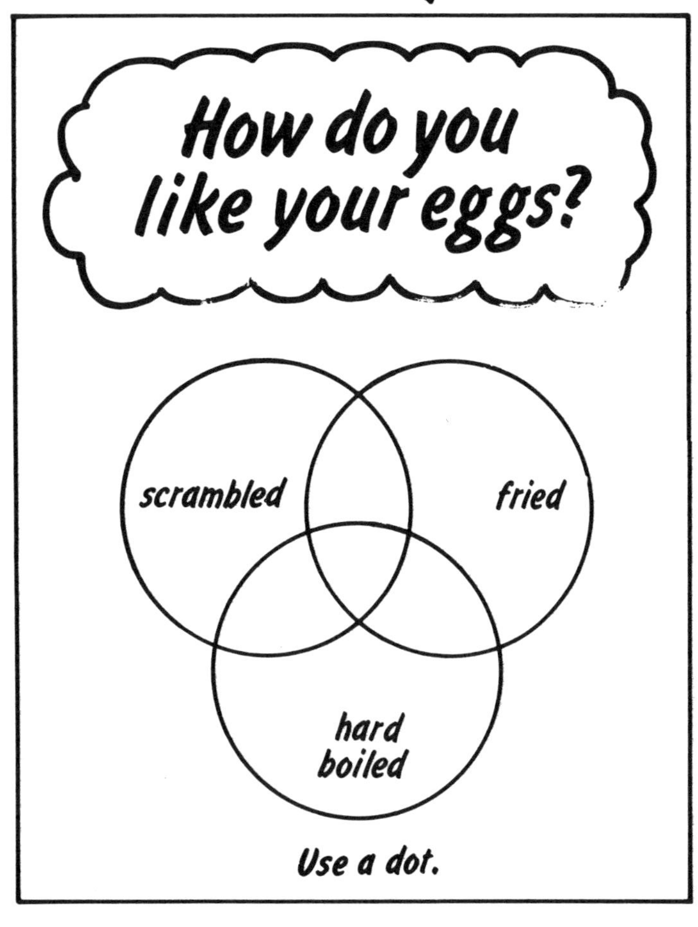

Do you usually eat breakfast?

yes ____________________

no ____________________

Use a tally mark.

What do you usually drink at breakfast?

juice ____________________

milk ____________________

coffee ____________________

soft drink ____________________

other ____________________

Use a tally mark.

You may choose more than one category.

Of these which do you like best?

pancakes ____________________

waffles ____________________

french toast ____________________

Use a tally mark.

Graphs that use Estimation

Estimate the length (in centimeters) of the line segment below.

4 ______

5 ______

6 ______

7 ______

8 ______

9 ______

more than 9 ______

Use a tally mark.

Cut a piece of string that you think is one foot long (without measuring). Tape your string on the line below.

too short	just about right	too long

About how many M&M's are in the jar?

Use a tally mark.

Less than 100 ______

100 to 200 ______

201 to 300 ______

301 to 400 ______

401 to 500 ______

501 to 600 ______

601 to 700 ______

more than 700 ______

How young do you think your teacher is?

20 - 25 ______

26 - 30 ______

31 - 35 ______

36 - 40 ______

41 - 45 ______

46 - 50 ______

over 50 ______

Use a tally mark.

Graphs about School

About how much time do you spend on homework each night?

less than an hour ____________________

about an hour ____________________

more than an hour ____________________

Use a tally mark.

Do you belong to any clubs or teams at school?

Yes	No

Use a dot.

Graphs about Sports and Exercise

On the average, how many days a week do you exercise?

(Do not count P.E. class)

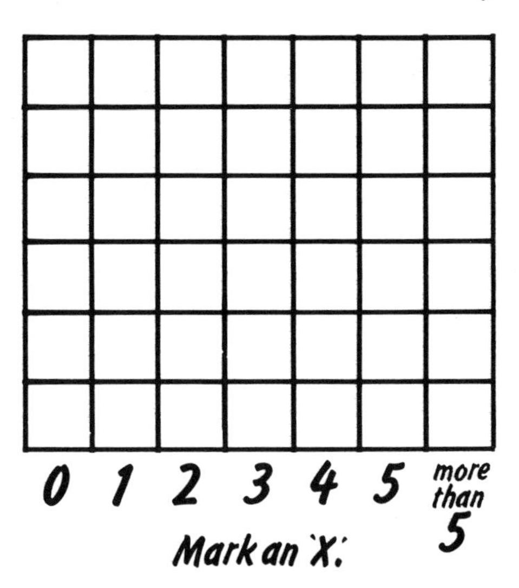

Mark an "X."

Which of these sports do you prefer to watch?

Soccer ____________

Football ____________

Tennis ____________

Volleyball ____________

Basketball ____________

Mark an "X."

Which of these sports do you prefer to play?

Soccer ____________

Football ____________

Tennis ____________

Volleyball ____________

Basketball ____________

Use a tally mark.

In which of these "cold" sports do you prefer to participate?

Snow skiing ____________

ice skating ____________

sledding ____________

ice hockey ____________

Use a tally mark.

Graphs about Entertainment

Which radio station do you listen to most often?

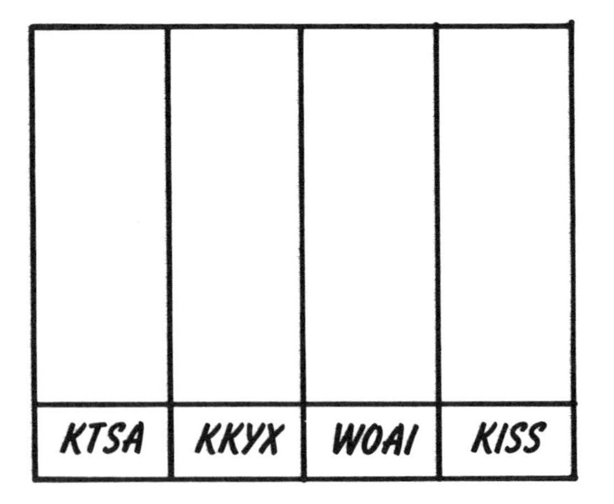

Use a dot.

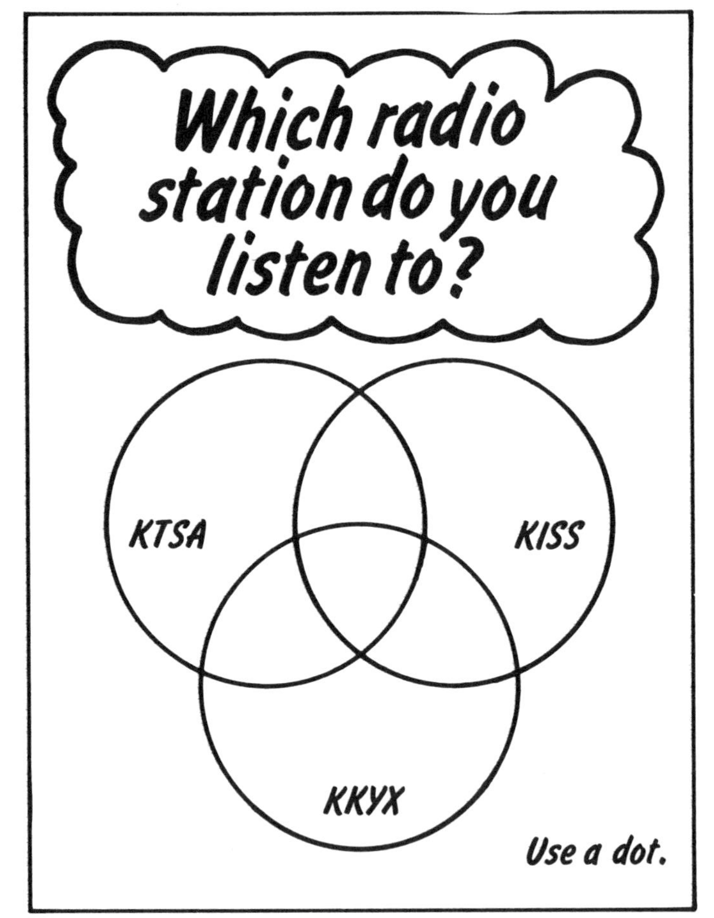

Use a dot.

How many hours a day (weekday) do you watch TV?

Use a tally mark.

Use a dot.

Note: As other movies become more popular, change the name.

More Graphs about Entertainment

Which would you prefer to visit?

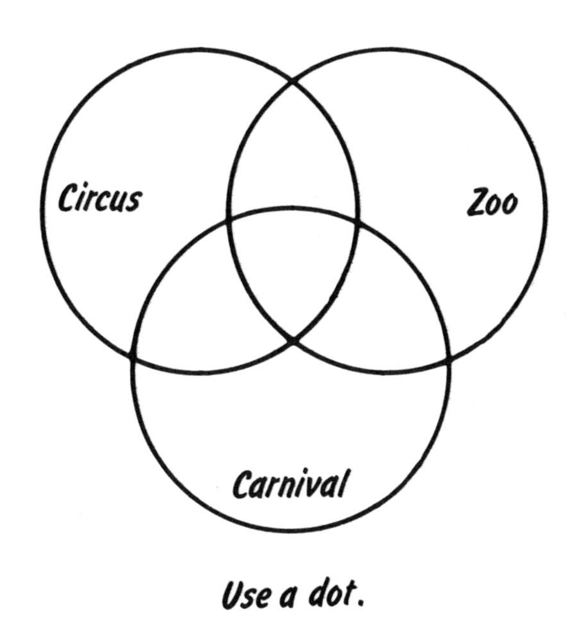

Use a dot.

Do you have a hobby?

Yes	No

Mark an 'X'.

Which type of movie do you prefer?

Horror ____________________

Musical ____________________

Science Fiction ____________________

Comedy ____________________

Romantic ____________________

Other ____________________

Use a tally mark.

Graphs about Food

Which chip do you prefer?

Frito ______________________

Potato Chip ______________________

Tortilla Chip ______________________

Chee-tos ______________________

Use a tally mark.

Which of these fruits do you like best?

Apple ______________________

Banana ______________________

Grapes ______________________

Oranges ______________________

Kiwi ______________________

Cantaloupe ______________________

Use a tally mark.

Do you like spinach?

Yes	No

Use a dot.

How do you like your steak cooked?

Rare ______________________

Medium Rare ______________________

Medium ______________________

Medium Well ______________________

Well ______________________

Use a tally mark.

More Graphs about Food

Chocolate chip ____________________

Peanut Butter ____________________

Sugar ____________________

Oatmeal ____________________

Use a tally mark.

Which after school snack do you like best?

Sandwich ____________________

Fruit ____________________

Cookies ____________________

Cake ____________________

Ice cream ____________________

chips ____________________

Use a tally mark.

What do you like on your hamburger?

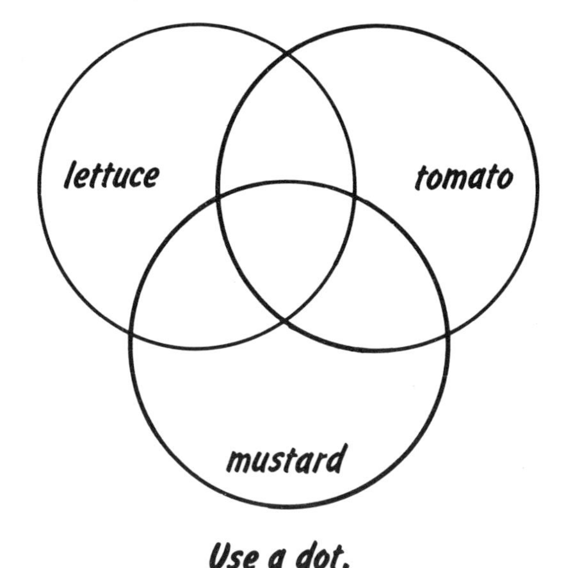

Use a dot.

Is there ice cream in your freezer?

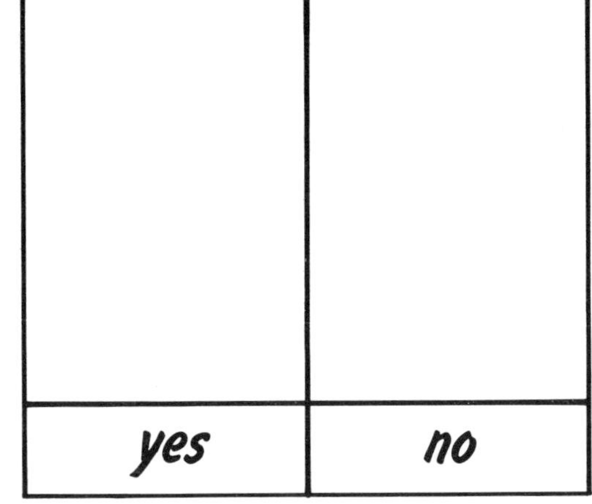

Use a dot.

Graphs about Home, Family and Friends

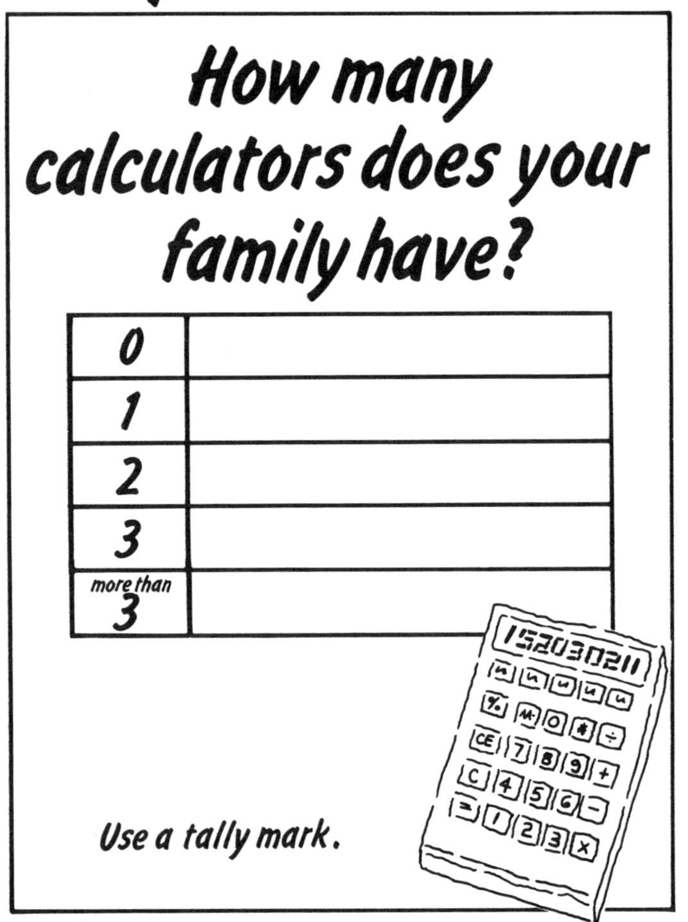

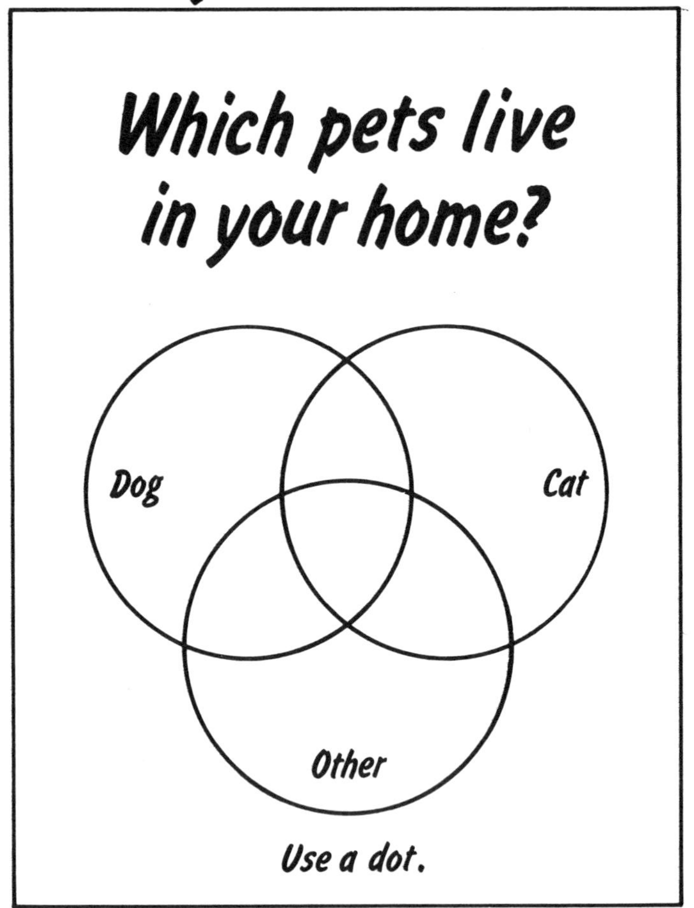

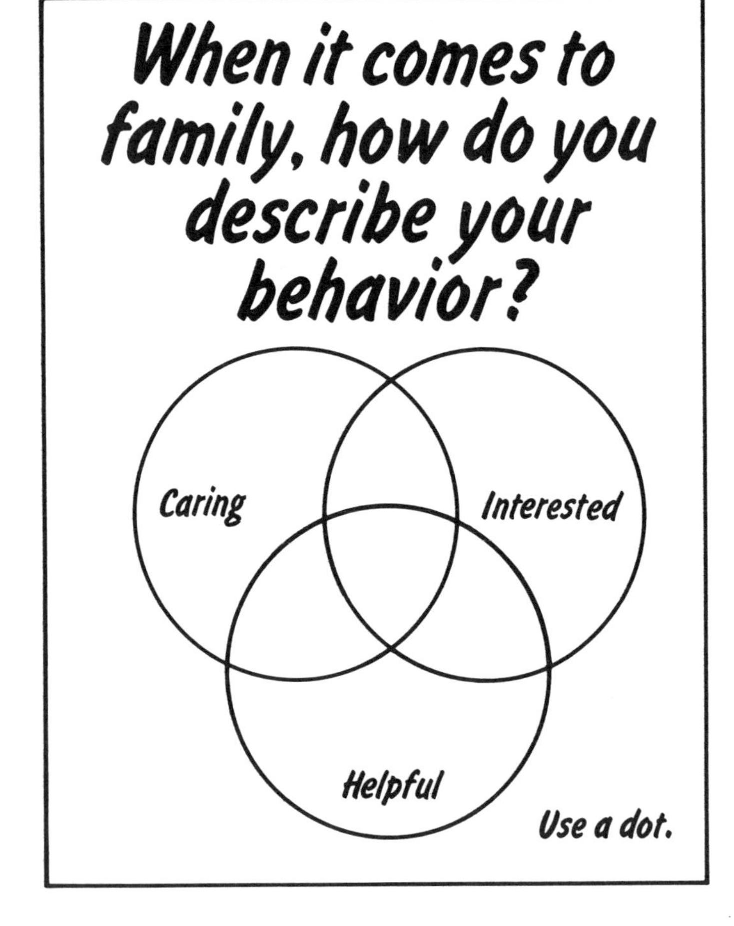

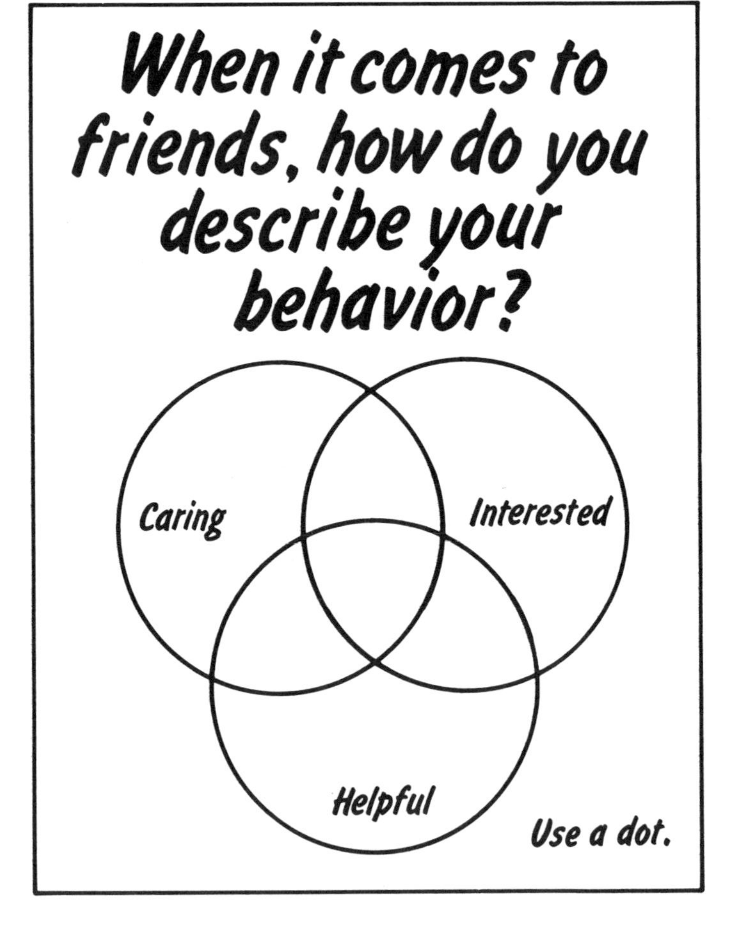

Graphs about Laws and Safety

Do you think there should be a law requiring the wearing of seat belts?

yes	no

Mark an 'X'.

Should motorcycle riders be required to wear helmets?

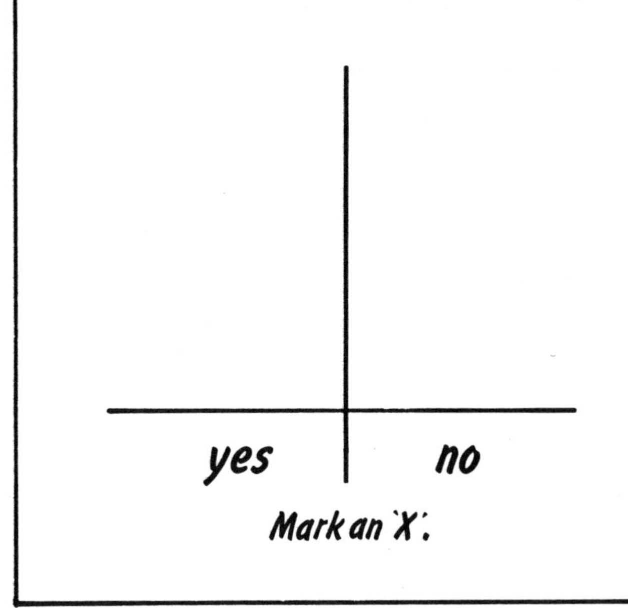

Mark an 'X'.

Have you ever had swimming lessons?

yes ______________________

no ______________________

Use a tally mark.

Have you ever had a first aid course?

yes ______________________

no ______________________

Use a tally mark.

Miscellaneous Topics

How many different ways can you double 47 mentally?

0 ____________________

1 ____________________

2 ____________________

3 ____________________

4 ____________________

more than 4 ____________________

Use a tally mark.

Where did you spend your summer?

San Antonio

other Texas city

outside of Texas

Use a dot.

Have you had a haircut in the past two months?

yes	no

Use a dot.

Which of these do you prefer to do on a Saturday afternoon?

movie ____________________

shop ____________________

visit relatives ____________________

stay home ____________________

Use a tally mark.

More Miscellaneous Topics

What color are your eyes?

Blue ______________________

Green ______________________

Brown ______________________

Hazel ______________________

Black ______________________

other ______________________

Use a tally mark.

Are you left-handed or right-handed?

right-handed	left-handed

Use a dot.

How much do you know about computers, and how do you feel about them?

	I know about computers	I don't know about computers
We don't need computers		
We can't live without computers		

Use a dot.

What is your favorite time of year?

Fall	Winter
Spring	Summer

Use a dot.

More Miscellaneous Topics

On what day of the month is your birthday?

1.______	13.______	25.______
2.______	14.______	26.______
3.______	15.______	27.______
4.______	16.______	28.______
5.______	17.______	29.______
6.______	18.______	30.______
7.______	19.______	31.______
8.______	20.______	
9.______	21.______	
10.______	22.______	
11.______	23.______	
12.______	24.______	

Use a tally mark.

Use a dot.

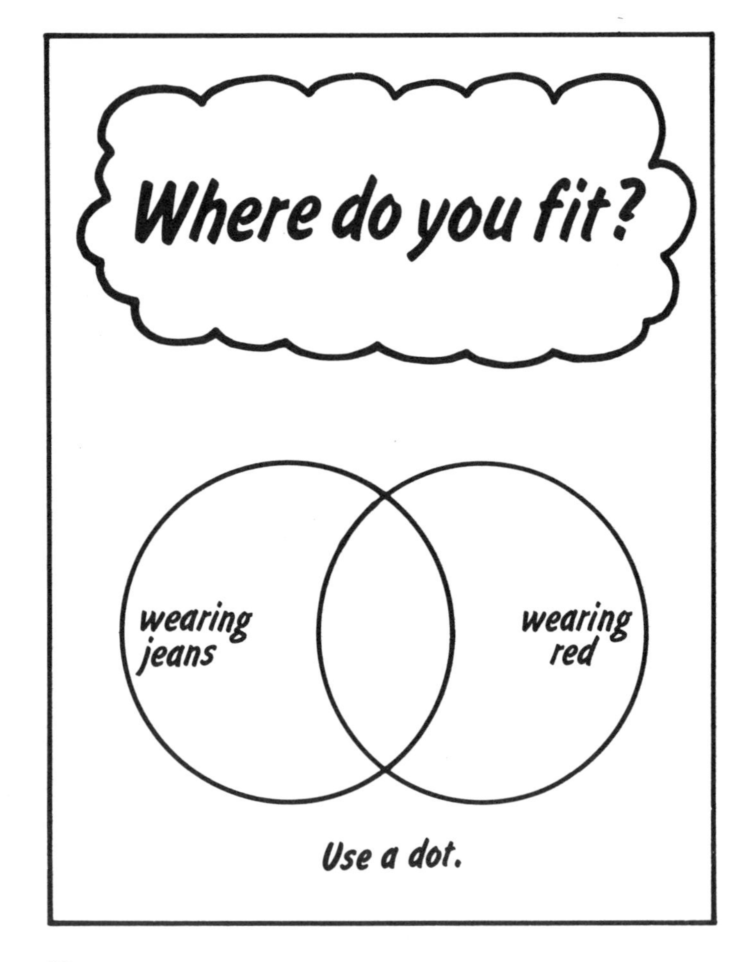

Use a dot.

Of these colors, which do you like best?

Red______

Yellow______

Green______

Blue______

Brown______

Orange______

Use a tally mark.